James Webb Space Telescope Mysteries

Copyright Page

This book is copyrighted for 2024

Title: James Webb Space Telescope Mysteries

Subtitle: What Existed Before the Big Bang?

The Living in Space Series Book 11

By Martin K. Ettington

All Rights Reserved USA 2024

ISBN: 9798335637374

Printed in the United States of America

James Webb Space Telescope Mysteries

James Webb Space Telescope Mysteries

The James Webb Space Telescope or JWST was launched on Christmas in 2021 to provide the most sensitive telescopic views of the Universe.

It has already revolutionized our real world views of the early Universe and these observations question our current theories of the creation of the Universe.

This book gets into the history of the JWST, the instruments it carries, new observations, and ideas of what these anomalies mean.

We review some of the most important observations made to date, and finally, some theories of how to explain these incredible views of the early reality of the Universe.

There are also some other anomalies in Space which are mentioned which were known before the JWST existed which also question the current Big Bang paradigm.

We are living in a revolutionary period of Astronomy much like the questions which occurred to Astronomers and Physicists in the early 20th Century which led to the theories we have today on Universe creation.

I'm sure that new paradigms will be the result of the current confusion which will contribute to new theories and better understandings of the world we live in.

James Webb Space Telescope Mysteries

James Webb Space Telescope Mysteries

Other books by Martin K. Ettington

Spiritual and Metaphysics Books:
Prophecy: A History and How to Guide
God Like Powers and Abilities
Enlightenment for Newbies
Removing Illusions to Find True Happiness
Using the Scientific Method to Study the Paranormal
A Compendium of Metaphysics and How to Guides (Six books together in one volume)
Love from the Heart
The Enlightenment Experience
Learn Your Soul's Purpose
Pursuing Enlightenment
A Modern Man's Search for Truth
Use Intuition and Prophecy to Improve Your Life
The Handbook of Spiritual and Energy Healing
Spiritual Masters

Longevity & Immortality:
Physical Immortality: A History and How to Guide
The Commentaries of Living Immortals
Records of Extremely Long Lived Persons
Enlightenment and Immortality
Longevity Improvements from Science
The 10 Principles of Personal Longevity
Telomeres & Longevity
The Diets and Lifestyles of the Worlds Oldest Peoples
The Longevity Six Books Bundle

Science Fiction:
The History of Science Fiction and Fantasy
Out of This Universe
Personal Freedom-Parts 1 & 2
The Psychic Soldier Series:
Book 1-Himalayan Journey
Book 2-A Soldier is Born
Book 3-Fighting For Right
Book 4-Earth Protector
The Immortality Sci Fi Bundle

The God Like Powers Series:
Human Invisibility
Invulnerability and Shielding
Teleportation
Psychokinesis
Our Energy Body, Auras, and Thoughtforms
The God Like Powers Series—Volume 1 Compilation
The Reality of Ghosts & Spirits

The Yoga Discovery Series:
Yoga-An Ancient Art Form
Hatha Yoga-Helping you Live Better
Raja Yoga-Through the Ages
The Yoga Discovery Package

Business & Coaching Books:
Creating, Paublishing, & Marketing Practitioner Ebooks
Building a Successful Longevity Coaching Business
Why Become a Coach?
The Professional Coaching Success Trilogy
2020-Make Money Writing and Selling Books
The 2020 Handbook of High Paying Work Without a College Degree

Science, Technology, and Misc.
Future Predictions By and Engineer & Seer
The Unusual Science & Technology Bundle
Removing Limits On Our Consciousness-And Thinking Outside the Box
Strange but True Stories and Facts
The Microscopic World Inside and Around Us
Radionics and Life Force Technologies
Infinity and Our Unbounded Universe
Planet Earth is Conscious: And Life Exists in Amazing Places
Accepted Science Which Is Likely Wrong
All About Ball Lightning
Stranger Than Science Stories and Facts
Stranger Than Science Stories and Facts (Book Two)
Quantum Mechanics, Technology, Consciousness, and the Multiverse
Universal Holistic Philosophy

Legendary Animals and Creatures
Are Cryptozoological Animals Real or Imaginary?
Fire in History and Mythology
All About Dragons
Sea Serpents and Ocean Monsters

James Webb Space Telescope Mysteries

The Importance of Fire in History and Mythology
Thunderbirds: Legends and Reality

<u>Ancient History</u>
The Real Atlantis-In the Eye of the Sahara
Ancient & Prehistoric Civilizations
Ancient & Prehistoric Civilizations-Book Two
The History of Antediluvian Giants
The Antediluvian History of Earth
Ancient Underground Cities and Tunnels
Strange Objects Which Should Not Exist
More Out of Place Artifacts
Strange and Ancient Places in the USA
A Theory of Ancient Prehistory And Giant Aliens
The Big Book of Pyramids Around the World
Underwater Ruins of Civilization

<u>Aliens and Space</u>
Aliens and Secret Technology
Aliens Are Already Among Us
Unidentified Submerged Objects and Underwater Bases
Four Evidences of Aliens and UFOs in Earth's History
Human And Alien UFO Anti-Gravity Research

<u>Living in Space</u>
Designing and Building Space Colonies
Humanity and the Universe
All About Moon Bases
All About Mars Journeys and Settlement
The Space and Aliens Six Books Bundle
A Theory of Ancient Prehistory and Giant Aliens
The Space Colonies and Space Structures Coloring Book
All About Asteroids
Spaceships, Past, Present, and Future
Astronauts, Cosmonauts, and Other Important Space Flyers
All About Mars Journeys and Settlement
Mining the Asteroid Belt
Exploring and Settling Our Huge Solar System

<u>Survival</u>
33 Incredible True Survival Stories
How to Survive Anything: From the Wilderness to Man Made Disasters
Building and Stocking a Nuclear Shelter for less than $10,000
Survival of Humanity Throughout the Ages

<u>Time Travel</u>
Real Time Travel Stories From a Psychic Engineer
The Real Nature of Time: An Analysis of Physics, Prophecy, and Time Travel Experiences

<u>Self Improvement</u>
The Importance of Genius in our World
Creating Your Own Reality
The Fear of Failure: And What You Can Do About It
Building Hope and Wonder Among Chaos.
A New Paradigm of Truth and Happiness
See The World Clearly: Be Happier and More Fulfilled
Stress Relief and Methods to Do So
The Importance of Creativity and How to Improve Yours Building Self Confidence

James Webb Space Telescope Mysteries

<u>The Longevity Training Series</u>

(A transcription of the online Multimedia Longevity Coaching Training Program)
The Personal Longevity Training Series-Book1-Long Lived Persons
The Personal Longevity Training Series-Book2-Your Soul's Purpose
The Personal Longevity Training Series-Book3-Enable Your Life Urge
The Personal Longevity Training Series-Book4-Your Spiritual Connection
The Personal Longevity Training Series-Book5-Having Love in Your Heart
The Personal Longevity Training Series-Book6-Energy Body Health
The Personal Longevity Training Series-Book7-The Science of Longevity
The Personal Longevity Training Series-Book8-Physical Body Health
The Personal Longevity Training Series-Book9-Avoiding Accidents
The Personal Longevity Training Series-Book10-Implementing These Principles
The Personal Longevity Training Series-Books One Thru Ten

These books are all available in digital, printed, and audio formats from my website and on Amazon, Barnes & Noble, Apple ITunes, and many other sites.

James Webb Space Telescope Mysteries

If you have any questions about this book or other subjects please contact the Author at:

mke@mkettingtonbooks.com

My Books Website is:

http://mkettingtonbooks.com

The Online Training Course site based on these books is:

https://www.mkettingtonbooks.com/training-home

Signup for our Monthly Newsletter and download past issues here:

https://personal-longevity.com/mkebooks/newsletter-signup-and-archive/

James Webb Space Telescope Mysteries

Table of Contents

1.0 Introduction 1
2.0 Accepted Theories of Cosmology 3
2.1 The Big Bang Theory 3
2.2 The Size of the Visible Universe 7
3.0 History of the JWST 11
4.0 JWST Scientific Instruments 15
5.0 Launch and Observation Location 25
6.0 Discoveries Made with the JWST 27
6.1 The Earliest Galaxies 27
6.2 Direct Images of Exoplanets 29
6.4 Oldest Black Hole 33
6.5 Oldest Proto Clusters of Galaxies 37
6.6 The Big Ring and Giant Arc 41
7.0 Other Space & Science Anomalies 47
8.0 Possible Explanations 61
8.1 The Universe Already Existed 61
8.2 Galaxies Formed Very Fast 65
8.3 We Really Live in a Multiverse 67
10.0 Bibliography 71

James Webb Space Telescope Mysteries

James Webb Space Telescope Mysteries

1.0 Introduction

The James Webb Space Telescope or JWST was launched on Christmas in 2021 to provide the most sensitive telescopic views of the Universe.

It has already revolutionized our real world views of the early Universe and these observations question our current theories of the creation of the Universe.

This book gets into the history of the JWST, the instruments it carries, new observations, and ideas of what these anomalies mean.

We review some of the most important observations made to date, and finally, some theories of how to explain these incredible views of the early reality of the Universe.

There are also some other anomalies in Space which are mentioned which were known before the JWST existed which also question the current Big Bang paradigm.

We are living in a revolutionary period of Astronomy much like the questions which occurred to Astronomers and Physicists in the early 20th Century which led to the theories we have today on Universe creation.

I'm sure that new paradigms will be the result of the current confusion which will contribute to new theories and better understandings of the world we live in.

James Webb Space Telescope Mysteries

James Webb Space Telescope Mysteries

2.0 Accepted Theories of Cosmology

Accepted theories of the Cosmology of the Universe which are now being questioned with new observations by the JWST. These theories are stated in this chapter:

2.1 The Big Bang Theory

The Big Bang is a physical theory that describes how the Universe expanded from an initial state of high density and temperature. It was first proposed as a physical theory in 1931 by Roman Catholic priest and physicist Georges Lemaître when he suggested the Universe emerged from a "primeval atom".

Various cosmological models of the Big Bang explain the evolution of the observable Universe from the earliest known periods through its subsequent large-scale form. These models offer a comprehensive explanation for a broad range of observed phenomena, including the abundance of light elements, the cosmic microwave background (CMB) radiation, and large-scale structure.

James Webb Space Telescope Mysteries

The uniformity of the Universe, known as the flatness problem, is explained through cosmic inflation: a sudden and very rapid expansion of space during the earliest moments.

Crucially, these models are compatible with the Hubble–Lemaître law—the observation that the farther away a galaxy is, the faster it is moving away from Earth. Extrapolating this cosmic expansion backward in time using the known laws of physics, the models describe an increasingly concentrated cosmos preceded by a singularity in which space and time lose meaning (typically named "the Big Bang singularity").

Physics lacks a widely accepted theory of quantum gravity that can model the earliest conditions of the Big Bang. In 1964 the CMB was discovered, which convinced many cosmologists that the competing steady-state model of cosmic evolution was falsified, since the Big Bang models predict a uniform background radiation caused by high temperatures and densities in the distant past.

A wide range of empirical evidence strongly favors the Big Bang event, which is now essentially universally accepted. Detailed measurements of the expansion rate of the Universe place the Big Bang singularity at an estimated 13.787 ± 0.020 billion years ago, which is considered the age of the Universe.

There remain aspects of the observed Universe that are not yet adequately explained by the Big Bang models. After its initial expansion, the Universe cooled sufficiently to allow the formation of subatomic particles, and later atoms. The unequal abundances of matter and antimatter that allowed this to occur is an unexplained effect known as baryon asymmetry. These primordial elements—mostly hydrogen, with some helium and lithium—later coalesced through gravity, forming early stars and galaxies.

Astronomers observe the gravitational effects of an unknown dark matter surrounding galaxies. Most of the gravitational potential in the Universe seems to be in this form, and the Big Bang models and various observations indicate that this excess gravitational potential is not created by baryonic matter, such as normal atoms. Measurements of the redshifts of supernovae indicate that the expansion of the Universe is accelerating, an observation attributed to an unexplained phenomenon known as dark energy.

James Webb Space Telescope Mysteries

2.2 The Size of the Visible Universe

The observable Universe is a sphere with a diameter of about 93 billion light-years, or 8.8×10^{26} meters. This is based on the estimated age of the Universe, 13.8 billion years, and the distance light travels in one year, 5.88 trillion miles.

Light from the farthest objects in the Universe has been traveling towards Earth for no more than 13.8 billion years, so the observable Universe is limited to that radius. However, the Universe is expanding, so the actual distance from Earth to the edge of the observable Universe may be closer to 46.5 billion light-years.

The observable Universe has a volume of about 410 nonillion cubic light-years, or 3.57×10^{80} cubic meters. For comparison, the Milky Way is about 100,000–

180,000 light-years in diameter, and the average distance between galaxies is 3 million light-years.

The Universe's total density and rate of expansion determine its geometry. Astronomers have measured these quantities and found that the Universe is likely "flat", which could mean it's infinite or has a very complex but finite geometry.

Cosmic Inflation

In physical cosmology, cosmic inflation, cosmological inflation, or just inflation, is a theory of exponential expansion of space in the early Universe. The inflationary epoch is believed to have lasted from 10^{-36} seconds to between 10^{-33} and 10^{-32} seconds after the Big Bang.

Following the inflationary period, the Universe continued to expand, but at a slower rate. The re-acceleration of this slowing expansion due to dark energy began after the Universe was already over 7.7 billion years old (5.4 billion years ago).

Inflation theory was developed in the late 1970s and early 80s, with notable contributions by several theoretical physicists, including Alexei Starobinsky at Landau Institute for Theoretical Physics, Alan Guth at Cornell University, and Andrei Linde at Lebedev Physical Institute. Alexei Starobinsky, Alan Guth, and Andrei Linde won the 2014 Kavli Prize "for pioneering the theory of cosmic inflation". It was developed further in the early 1980s. It explains the origin of the large-scale structure of the cosmos. Quantum fluctuations in the microscopic inflationary region, magnified to cosmic size, become the seeds for the growth of structure in the Universe. Many physicists also believe that inflation explains why the Universe appears to be the same in all directions (isotropic), why the cosmic

James Webb Space Telescope Mysteries

microwave background radiation is distributed evenly, why the Universe is flat, and why no magnetic monopoles have been observed.

The detailed particle physics mechanism responsible for inflation is unknown. The basic inflationary paradigm is accepted by most physicists, as a number of inflation model predictions have been confirmed by observation; however, a substantial minority of scientists dissent from this position. The hypothetical field thought to be responsible for inflation is called the inflation.

James Webb Space Telescope Mysteries

James Webb Space Telescope Mysteries

3.0 History of the JWST

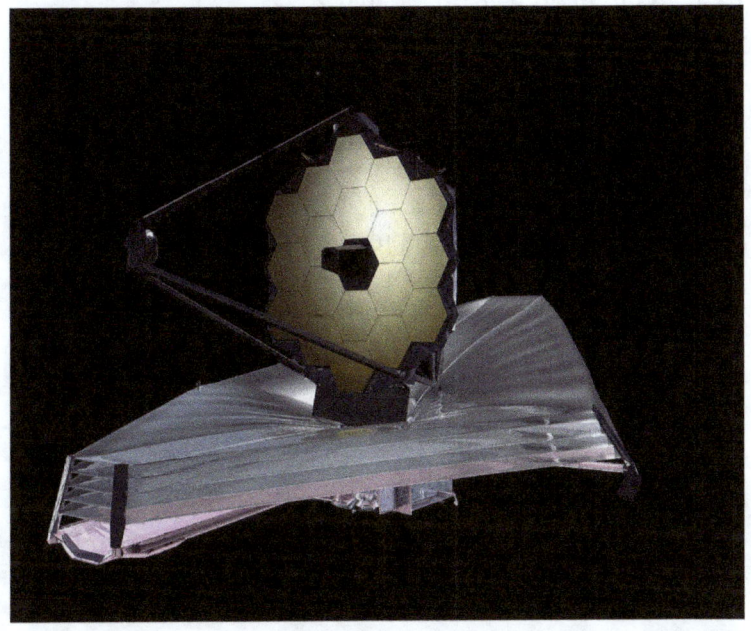

The James Webb Space Telescope (JWST) is a space telescope designed to conduct infrared astronomy. As the largest telescope in space, it is equipped with high-resolution and high-sensitivity instruments, allowing it to view objects too old, distant, or faint for the Hubble Space Telescope. This enables investigations across many fields of astronomy and cosmology, such as observation of the first stars and the formation of the first galaxies, and detailed atmospheric characterization of potentially habitable exoplanets.

The Webb was launched on 25 December 2021 on an Ariane 5 rocket from Kourou, French Guiana. In January 2022 it arrived at its destination, a solar orbit near the Sun–Earth L2 Lagrange point, about 1.5 million kilometers

James Webb Space Telescope Mysteries

(930,000 mi) from Earth. The telescope's first image was released to the public on 11 July 2022.

The U.S. National Aeronautics and Space Administration (NASA) led Webb's design and development and partnered with two main agencies: the European Space Agency (ESA) and the Canadian Space Agency (CSA). The NASA Goddard Space Flight Center in Maryland managed telescope development, while the Space Telescope Science Institute in Baltimore on the Homewood Campus of Johns Hopkins University operates Webb. The primary contractor for the project was Northrop Grumman.

The telescope is named after James E. Webb, who was the administrator of NASA from 1961 to 1968 during the Mercury, Gemini, and Apollo programs.

Webb's primary mirror consists of 18 hexagonal mirror segments made of gold-plated beryllium, which together create a 6.5-meter-diameter (21 ft) mirror, compared with Hubble's 2.4 m (7 ft 10 in). This gives Webb a light-collecting area of about 25 m2 (270 sq. ft), about six times that of Hubble. Unlike Hubble, which observes in the near ultraviolet and visible (0.1 to 0.8 μm), and near infrared (0.8–2.5 μm) spectra, Webb observes a lower frequency range, from long-wavelength visible light (red) through mid-infrared (0.6–28.3 μm). The telescope must be kept extremely cold, below 50 K (−223 °C; −370 °F), so that the infrared light emitted by the telescope itself does not interfere with the collected light. Its five-layer sunshield protects it from warming by the Sun, Earth, and Moon.

Initial designs for the telescope, then named the Next Generation Space Telescope, began in 1996. Two concept studies were commissioned in 1999, for a potential launch in 2007 and a US$1 billion budget. The program was

James Webb Space Telescope Mysteries

plagued with enormous cost overruns and delays. A major redesign was accomplished in 2005, with construction completed in 2016, followed by years of exhaustive testing, at a total cost of US$10 billion.

James Webb Space Telescope Mysteries

James Webb Space Telescope Mysteries

4.0 JWST Scientific Instruments

There are a variety of very sensitive instruments on the JWST. These instruments are critical to have for the sensitivity needed to make new discoveries. Here are the main instruments:

Webb's unprecedented scientific power is a function of both the size of its primary mirror and the extreme sensitivity and precision of its four scientific instruments:

- Mid-Infrared Instrument (MIRI)
- Near-Infrared Camera (NIRCam)
- Near-Infrared Spectrograph (NIRSpec)
- Near-Infrared Imager and Slitless Spectrograph/Fine Guidance Sensor (NIRISS/FGS)

All scientific journal articles and many press releases will refer to specific instruments, instrument components, or observing modes used for observations with Webb.

This section provides some clarity on the function of each component, observing mode, and instrument, and the types of observations they are designed for.

About Webb's Instruments-Overview

Unlike simple backyard telescopes, which focus light from space directly into the eye, research telescopes include scientific instruments that record light precisely. Scientific instruments are crucial elements in both ground- and space-based telescopes, and are designed to optimize observations for scientific use.

During an observation with Webb, infrared light travelling from the target object or region of space is intercepted by the primary mirror, reflected onto the smaller secondary

James Webb Space Telescope Mysteries

mirror, and then focused into the Integrated Science Instrument Module (ISIM). Mirrors then direct the light into one or more of the four scientific instruments, which may focus, filter, block, or disperse the light before it is recorded.

Instrument Components

Each of Webb's four instruments includes a set of components that are common in research telescopes, along with components that are more specialized.

Common Components

Cameras capture two-dimensional images of regions of space. NIRCam and NIRISS capture images in the near-infrared, while MIRI captures mid-infrared images. NIRSpec is the only instrument without a camera.

Spectrographs spread light out into a spectrum so that the brightness of each individual wavelength can be measured. Webb has a number of different types of spectrographs, each designed for a slightly different purpose. All four of Webb's instruments have spectrographs.

Coronagraphs are opaque disks used to block the bright light of stars in order to detect the much fainter light of planets and debris disks orbiting the star. NIRCam and MIRI have coronagraphs.

Filters are thin sheets of specialized materials designed to transmit a certain range of wavelengths of light and block all others. Webb's filters are similar to light filters on handheld cameras, and are used in conjunction with cameras, coronagraphs, and spectrographs. All four

James Webb Space Telescope Mysteries

instruments include numerous filters, including broad-band filters, which transmit a wide range of wavelengths; narrow-band filters, which transmit a very narrow range of wavelengths; and clear filters, which transmit all wavelengths collected by the telescope.

Detectors absorb light and convert it into electrical charges so that the information carried in the light (brightness, wavelength, and position) can be stored as digital data before being converted into radio signals and transmitted to Earth. Detectors are arranged in arrays, and are equivalent to the CCDs in a digital camera or the film in an analog camera. All four instruments have at least two detectors. NIRSpec, NIRISS, and NIRCam detectors are sensitive to near-infrared light (0.6 – 4.9 µm). MIRI detectors are sensitive to mid-infrared light (4.9 – 28.8 microns).

Specialized Components

Webb's **microshutter array** (MSA) is a grid of 248,000 tiny doors that can be opened and closed to transmit or block light in order to capture spectra of 100 individual objects or points in space at the same time (multi-object spectroscopy). NIRSpec is the only instrument with an MSA, and Webb is the only space telescope with an MSA.

The **integral field unit** (IFU) is a combination of camera and spectrograph used to capture and map spectra across a field of view in order to understand variation over space. NIRSpec and MIRI have IFUs.

Webb's **aperture mask** is a metal plate with seven hexagonal holes that is placed in front of the detectors to increase the effective resolution of the telescope and capture more detailed images of extremely bright objects

James Webb Space Telescope Mysteries

(aperture mask interferometry). NIRISS is the only instrument with an aperture mask.

Observing Modes-Instrument Wavelength Range, Field of View, and Resolution

Instrument Detail

Each of Webb's four instruments is designed to study a wide range of objects and phenomena in space, including planets, stars, galaxies, gas clouds, debris disks, black holes, and dark matter.

What makes each instrument unique is its specific combination of components, observing modes, wavelength range, field of view, and resolution.

While some investigations are conducted with a single instrument and observing mode, most rely on a combination of instruments and/or observing modes.

Mid-Infrared Instrument (MIRI)

James Webb Space Telescope Mysteries

Components: Camera, Coronagraphs, Spectrographs, Integral Field Unit

Wavelength range: 4.9 µm – 27.9 µm (mid-infrared, which is unique to MIRI)

Detectors: Arsenic-doped silicon

Imaging modes: Standard Imaging, Coronographic Imaging, Time-Series Imaging

Spectroscopy modes: Single-Object Slitless Spectroscopy, Slit Spectroscopy, Integral Field Unit Spectroscopy, Time-Series Spectroscopy

Resolution: Medium-resolution imaging; Low- and medium-resolution spectroscopy

MIRI provides imaging and spectroscopy capabilities in the mid-infrared. As the only mid-infrared instrument, astronomers rely on MIRI to study cooler objects like debris disks, which emit most of their light in the mid-infrared, and extremely distant galaxies whose light has been shifted into the mid-infrared over time.
MIRI was developed through a collaboration between the European Consortium (EC) and the Jet Propulsion Laboratory (JPL).

James Webb Space Telescope Mysteries

Near-Infrared Camera (NIRCam)

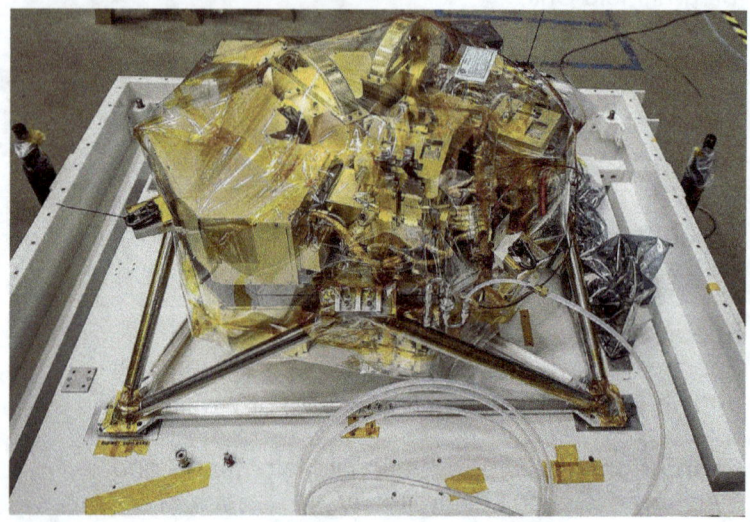

Components: Camera, Coronagraphs, Spectrographs
Wavelength range: 0.6 μm – 5 μm (red to near-infrared)
Detectors: Mercury cadmium telluride
Imaging modes Standard Imaging, Coronagraphic Imaging, Time-Series Imaging

Spectroscopy modes: Wide-Field Slitless Spectroscopy, Time-Series Spectroscopy

Resolution: High-resolution imaging and spectroscopy
NIRCam is Webb's primary near-infrared imager, providing high-resolution imaging and spectroscopy for a wide variety of investigations.

Because NIRCam is the only near-infrared instrument with coronagraphic and time-series imaging capabilities, it is crucial for many exoplanet studies.

James Webb Space Telescope Mysteries

In addition to imaging and spectroscopy, NIRCam is also part of Webb's wavefront sensing and control system, which detects and corrects for slight irregularities in the shape of the primary mirror or misalignment between mirror segments, giving the telescope the ability to focus clearly on objects near and far.

NIRCam was built by a team at the University of Arizona and Lockheed Martin's Advanced Technology Center.

Near-Infrared Spectrograph (NIRSpec)

Components: Spectrographs, Integral Field Unit, Microshutter Array (Unique to NIRSpec)

Wavelength range: 0.6 μm – 5.3 μm (red to near-infrared)

Detectors: Mercury cadmium telluride

James Webb Space Telescope Mysteries

Imaging modes: N/A (with the exception of images collected during Integral Field Unit Spectroscopy)

Spectroscopy modes: Slit Spectroscopy, Multi-Object Spectroscopy (Unique to NIRSpec), Integral Field Unit Spectroscopy, Time-Series Spectroscopy

Resolution: Low-, Medium-, and High-resolution spectroscopy

NIRSpec is one of Webb's versatile tools for near-infrared spectroscopy. In addition to standard single-slit spectroscopy to gather spectra of specific objects, NIRSpec also has an integral field unit to investigate spatial variations in spectra and a microshutter array to capture individual spectra of dozens of objects at once.

This highly efficient design is part of what makes Webb ideal for studying extremely distant, faint galaxies.

NIRSpec was built for the European Space Agency by Airbus Industries with the microshutter array (MSA) and detector sub-systems fabricated by NASA.

James Webb Space Telescope Mysteries

Near-Infrared Imager and Slitless Spectrograph (NIRISS)/Fine Guidance Sensor (FGS)

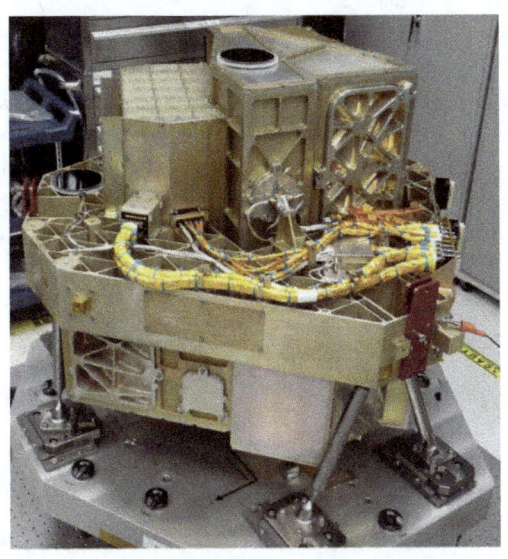

Components: Camera, Spectrographs, Aperture Mask

Wavelength range: 0.6 μm – 5 μm (red to near-infrared)

Detectors: Mercury cadmium telluride

Imaging modes: Standard Imaging, Aperture Mask Interferometry (Unique to NIRISS)

Spectroscopy modes: Wide-Field Slitless Spectroscopy, Single-Slit Spectroscopy

Resolution: High-resolution imaging; Low- and Medium-resolution spectroscopy

NIRISS provides near-infrared imaging and spectroscopic capabilities. As the only instrument capable of aperture

James Webb Space Telescope Mysteries

mask interferometry, NIRISS has the unique ability to capture images of bright objects at a resolution greater than the other imagers.

Housed in the same assembly as NIRISS is Webb's Fine Guidance Sensor (FGS). The FGS is a camera system designed to make sure Webb is stable and pointing in exactly the right direction throughout the observation. The FGS detects and identifies guide stars and ensures that Webb is locked onto those stars for the entire observation.

James Webb Space Telescope Mysteries

5.0 Launch and Observation Location

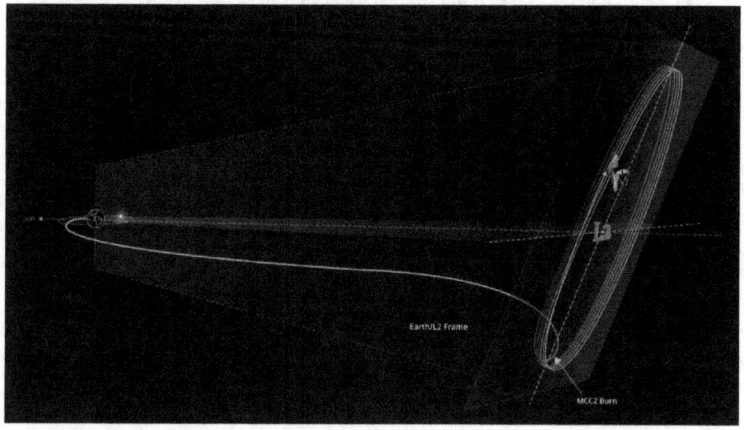

Webb operates in a halo orbit, circling around a point in space known as the Sun–Earth L2 Lagrange point, approximately 1,500,000 km (930,000 mi) beyond Earth's orbit around the Sun. Its actual position varies between about 250,000 and 832,000 km (155,000–517,000 mi) from L2 as it orbits, keeping it out of both Earth and Moon's shadow. By way of comparison, Hubble orbits 550 km (340 mi) above Earth's surface, and the Moon is roughly 400,000 km (250,000 mi) from Earth.

Objects near this Sun–Earth L2 point can orbit the Sun in synchrony with the Earth, allowing the telescope to remain at a roughly constant distance with continuous orientation of its sunshield and equipment bus toward the Sun, Earth and Moon. Combined with its wide shadow-avoiding orbit, the telescope can simultaneously block incoming heat and light from all three of these bodies and avoid even the smallest changes of temperature from Earth and Moon shadows that would affect the structure, yet still maintain uninterrupted solar power and Earth communications on its sun-facing side. This arrangement keeps the temperature of the spacecraft constant and below the 50 K (−223 °C; −370 °F) necessary for faint infrared observations.

James Webb Space Telescope Mysteries

6.0 Discoveries Made with the JWST

The JWST has quickly made some incredible discoveries which are blowing away the current theories of the creations of the Universe. The "Big Bang" Theory doesn't look so solid these days.

Additional discoveries about Exo-Planets may not break scientific theories but they are important since they tell us about additional potential life in our Galaxy.

6.1 The Earliest Galaxies

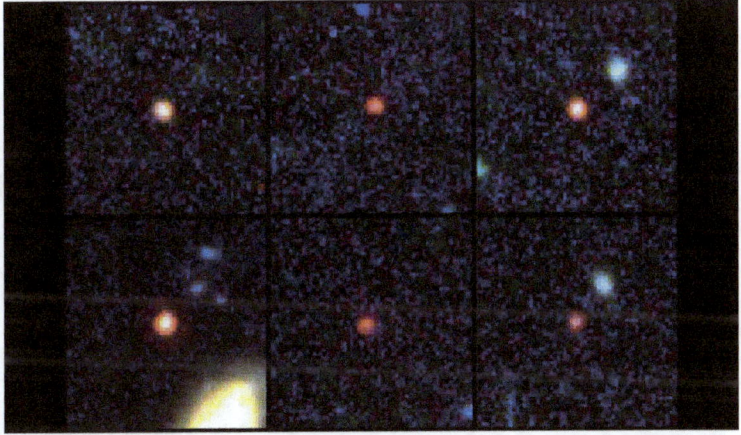

The JWST was made to observe the most distant galaxies in the Universe, and in mid-December, scientists confirmed that they had done just that. The telescope has officially observed the four most distant galaxies known, which also means they are the oldest. Webb observed the galaxies as they appeared about 13.4 billion years ago, when the Universe was only 350 million years old, about 2% of its current age.

James Webb Space Telescope Mysteries

Scientists suspected that the four galaxies were incredibly ancient, like hundreds of others identified by Webb. As part of the JWST Advanced Deep Extragalactic Survey.

(JADES) researchers confirmed their age, analyzing data from the telescope's Near Infrared Spectrograph to find out how fast the galaxies were moving away from the telescope. This is the galaxies' redshift — how much the wavelengths of light they shed have lengthened as the Universe expands. Their redshift was 13.2, the highest ever measured.

"These [galaxies] are well beyond what we could have imagined finding before JWST," Brant Robertson, an astrophysicist at the University California Santa Cruz and one of the researchers involved in the observations, said in a statement. "With JWST, for the first time we can now find such distant galaxies and then confirm spectroscopically that they really are that far away."

James Webb Space Telescope Mysteries

6.2 Direct Images of Exoplanets

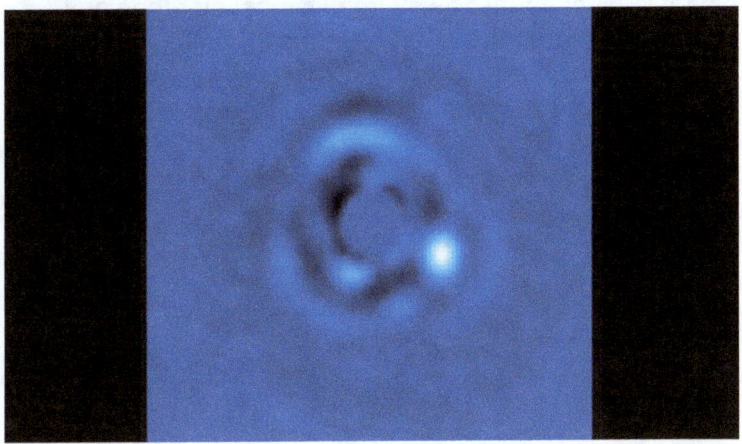

Scientists discovered the first exoplanets in the 1990s, and today there are over 3,000 known worlds orbiting faraway stars. Still, only around two dozen of these have been imaged directly. Most exoplanets are so far away that they can only be detected through a dip in the light of the star they are orbiting, when that planet passes in front of its host star. But Webb could change that. In September, it captured its first direct image of an exoplanet.

"This is a transformative moment, not only for Webb but also for astronomy generally," Sasha Hinkley, an astronomer at the University of Exeter in the U.K. who led these observations, said in a statement in September.

The planet, called HIP 65426 b, was discovered in 2017. To view it, scientists used two of Webb's cameras, several filters, and the telescope's coronagraphs, tools which blocked out the light of the central star. Along with the telescope's exceptional sensitivity, the planet has several features that make it easier to observe.

James Webb Space Telescope Mysteries

At 100 times the distance from our sun to the Earth, this planet is much farther away from its host star than any planet in our solar system (in contrast, Pluto is only 40 times that sun-Earth distance from our sun). A colossal gas giant, it's also exceptionally large ☐ — about 12 times the size of Jupiter.

James Webb Space Telescope Mysteries

6.3 Measurements of Exo-Planet Atmospheres

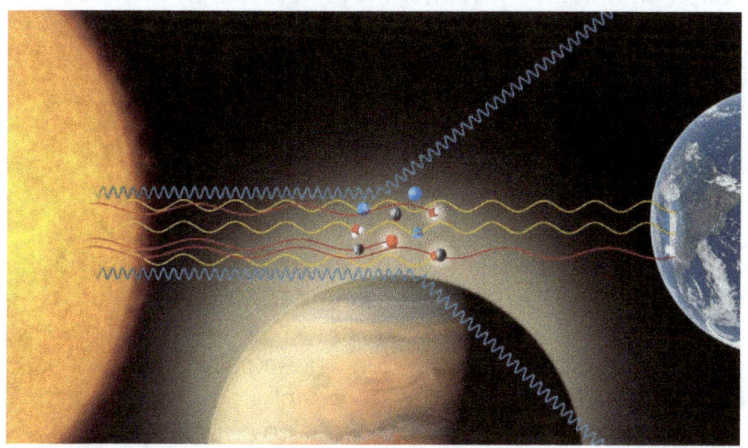

Current measurements of the atmospheres of Exo-Planets have led to discoveries of Methane and other chemicals which are indicative of life existing on these planets.

Recent reports of the James Webb Space Telescope (JWST) detecting signs of life of a distant planet outside the solar system are, unfortunately, somewhat premature. That's the conclusion of research conducted by scientists from the University of California Riverside (UCR).

While likely to disappoint all of us eager for the confirmation of extraterrestrial life, however, it doesn't mean the JWST won't find traces of life in the atmosphere of an extrasolar planet, or "exoplanet," in the future.

The recent excitement around the potential detection of life signs on an exoplanet started in 2023 when the JWST detected potential "biosignature" elements in the atmosphere of the exoplanet K2-18 b, a super-Earth located around 120 light-years from Earth.

James Webb Space Telescope Mysteries

Though many exoplanets are extreme, violent or at least "alien" in nature — whether they're blasted by intense radiation from their stars, lack a solid surface or are frozen relics at the edge of their systems — K2-18 b was a tantalizing target in the search for life because it is rather similar to our planet.

An Earth-like ocean world

K2-18 b is between two and three times the width of Earth with 8.6 times the mass of our planet. It's also located in the habitable zone of its star, the region neither too hot nor too cold to support liquid water. The exoplanet is thus theorized to be an ocean, or "hycean" world, replete with liquid water — a vital ingredient for life as we know it.

Unlike Earth, however, the atmosphere of this exoplanet seems to be mainly hydrogen rather than nitrogen.

"This planet gets almost the same amount of solar radiation as Earth. And if atmosphere is removed as a factor, K2-18 b has a temperature close to Earth's, which is also an ideal situation in which to find life," team member and UCR project scientist Shang-Min Tsai said in a statement.

6.4 Oldest Black Hole

The James Webb Space Telescope (JWST) has spotted the oldest black hole ever seen, an ancient monster with the mass of 1.6 million suns lurking 13 billion years in the Universe's past. This is considered too old to have formed with current theories about Black Hole Formation.

The James Webb Space Telescope, whose cameras enable it to look back in time to our Universe's beginnings, spotted the supermassive black hole at the center of the infant galaxy GN-z11 just 440 million years after the Universe began.

And the space-time rupture isn't alone, it's one of countless black holes that gorged themselves to terrifying scales during the cosmic dawn — the period about 100 million years after the Big Bang when the young Universe began glowing for a billion years.

James Webb Space Telescope Mysteries

How the cosmic whirlpools ballooned in scale so rapidly after the Universe began isn't clear. But looking for an answer could help explain how today's supermassive black holes — which anchor entire galaxies including our Milky Way — grew to such mind-boggling sizes. The researchers published their findings Jan. 17 in the journal Nature.

Black holes in the early Universe "can't grow quietly and gently as many black holes do in the local [present-day] Universe," lead-author Roberto Maiolino, a professor of astrophysics at the University of Cambridge, told Live Science. "They must experience some peculiar birth or formation, and some peculiar growth."

Closer to the present-day, astronomers believe black holes are born from the collapse of giant stars. But however they come to be, they grow by ceaselessly gorging on gas, dust, stars and other black holes. As they feast, friction causes the material spiraling into the black holes' maws to heat up, and they emit light that can be detected by telescopes — turning them into so-called active galactic nuclei (AGN).

The most extreme AGN are quasars, supermassive black holes that are billions of times heavier than the sun and shed their gaseous cocoons with light blasts trillions of times more luminous than the brightest stars.

Because light travels at a fixed speed through the vacuum of space, the deeper that scientists look into the Universe, the more remote light they intercept and the further back in time they see.

To spot the black hole in the new study, the astronomers scanned the sky with two infrared cameras — the JWST's Mid-Infrared Instrument (MIRI) and Near Infrared Camera

James Webb Space Telescope Mysteries

— and used the cameras' built-in spectrographs to break down the light into its component frequencies.

By deconstructing these faint glimmers from the Universe's earliest years, they found an unexpected spike among the frequencies contained within the light — a key sign that the hot material around a black hole was beaming out faint traces of light across the Universe.

The most popular explanations for how these early black holes grew so fast are that they formed from the sudden collapse of giant gas clouds or that they came from many mergers between clumps of stars and black holes.

Nonetheless, astronomers haven't ruled out that some of these black holes could have been seeded by hypothesized "primordial" black holes, thought to be created moments after — and in some theories even before — the Universe began.

"It's not so clear that [direct collapse] is the only way to make a black hole, because you need some special circumstances for it to happen," Maiolino said. "You need it to be a pristine cloud, yet to be enriched by heavy elements made by the first stars, and one that is fairly massive — from 10,000 to up to a million solar masses."

To prevent such a cloud from cooling too quickly and collapsing into massive stars first, it must also be beamed with ultraviolet light, likely from a nearby galaxy or black hole.

"So you need this peculiar condition where the cloud is not getting enriched [by absorbing exploded star material], but is also next to another galaxy which is producing a lot of photons," Maiolino said. "So we're not necessarily looking for a single scenario, actually two or more of them could be at play."

6.5 Oldest Proto Clusters of Galaxies

Astronomers using the NASA/ESA/CSA James Webb Space Telescope have spotted at least five young globular clusters within SPT 0615-JD1 (also known as the Cosmic Gems arc), a strongly-lensed galaxy that existed when the Universe was 460 million years old.

These images show the galaxy cluster SPT-CL J0615-5746 (right) and a portion of this cluster (left) showing two distinct lensed galaxies. The Cosmic Gems arc is shown with several galaxy clusters.

"These galaxies are thought to be a prime source of the intense radiation that reionized the early Universe," said Dr. Angela Adamo, an astronomer at Stockholm University and the Oskar Klein Centre.

"What is special about the Cosmic Gems arc is that thanks to gravitational lensing we can actually resolve the galaxy down to parsec scales."

James Webb Space Telescope Mysteries

SPT 0615-JD1 was initially discovered in Hubble images obtained by the RELICS (Reionization Lensing Cluster Survey) program of the lensing galaxy cluster SPT-CL J0615-5746, which is located approximately 7.7 billion light-years away in the constellation of Pictor.

With Webb, Dr. Adamo and colleagues can now see where stars formed and how they are distributed, in a similar way to how Hubble is used to study local galaxies.

Webb's view provides a unique opportunity to study star formation and the inner workings of infant galaxies at such an unprecedented distance.

"Webb's incredible sensitivity and angular resolution at near-infrared wavelengths, combined with gravitational lensing provided by the massive foreground galaxy cluster, enabled this discovery. No other telescope could have made this discovery," said Dr. Larry Bradley, an astronomer at the Space Telescope Science Institute.

"The surprise and astonishment was incredible when we opened the Webb images for the first time," Dr. Adamo said.

"We saw a little chain of bright dots, mirrored from one side to the other — these cosmic gems are star clusters." "Without Webb we would not have known we were looking at star clusters near such a young galaxy."

The astronomers note that their discovery connects a variety of scientific fields.

"These results provide direct evidence that indicates proto-globular clusters formed in faint galaxies during the

James Webb Space Telescope Mysteries

reionization era, which contributes to our understanding of how these galaxies have succeeded in reionizing the Universe," Dr. Adamo said.

"This discovery also places important constraints on the formation of globular clusters and their initial properties."

"For instance, the high stellar densities found in the clusters provide us with the first indication of the processes taking place in their interiors, giving new insights into the possible formation of very massive stars and black hole seeds, which are both important for galaxy evolution."

In the future, the team hopes to build a sample of galaxies for which similar resolutions can be achieved.

"I am confident there are other systems like this waiting to be uncovered in the early Universe, enabling us to further our understanding of early galaxies," said Dr. Eros Vanzella, an astronomer at the INAF – Astrophysics and Space Science Observatory of Bologna.

James Webb Space Telescope Mysteries

James Webb Space Telescope Mysteries

6.6 The Big Ring and Giant Arc

An impossibly huge ring of galaxies might lead us to new physics. Here's how: The ring of galaxies, appropriately called the "Big Ring," has a circumference of nearly 4 billion light-years.

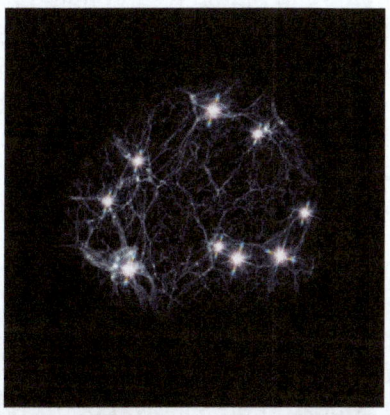

The location of the Big Ring (in blue) and the Giant Arc (in red) on the sky.

James Webb Space Telescope Mysteries

An intergalactic ring-shaped superstructure of galaxies and galaxy clusters — so large it defies explanation — has been discovered. This is a structure that lives so deep in the Universe that we see it as it was some 9.2 billion years ago.

The huge superstructure, nicknamed the "Big Ring," spans 1.3 billion light-years in diameter and has a circumference of about 4 billion light-years. It is also close to another immense superstructure, the "Giant Arc in the Sky," which is actually even larger with a diameter of 3.3 billion light-years. The Giant Arc sits at a similar distance to us in the constellation of Bootes, the Herdsman. Alas, these superstructures are far too faint to be seen with a backyard telescope.

In fact, both superstructures were discovered in observations performed by the 2.5-meter telescope of the Sloan Digital Sky Survey at Apache Point in New Mexico, U.S. by Alexia Lopez. Interestingly, Lopez, a Ph.D. student at the University of Central Lancashire in the U.K., spotted the galaxies in these superstructures not because they are bright, but rather because they absorb some of the light emanating from more distant quasars. Quasars are the extremely luminous interiors of active galaxies; they're powered by supermassive black holes.

"Identifying two extraordinary ultra-large structures in such close configuration raises the possibility that together they form an even more extraordinary cosmological system," said Lopez in a statement.

The Big Ring isn't actually even a ring – it's coiled sort of like a slinky. Plus, we see it edge-on.

James Webb Space Telescope Mysteries

Still, the problem with the Big Ring and the Giant Arc (and other similar superstructures, for that matter) is they defy cosmological theory.

According to theory, all structures in the Universe can be traced back to what's known as the cosmic microwave background (CMB) radiation — the so-called "fireball of the Big Bang" that scientists observe filling the Universe.

During the first 300,000 years of cosmic history, the Universe was a sea of dense plasma — that is, atomic nuclei and free electrons. Waves crashed through this plasma, with matter bunching up at the peaks and becoming more sparse in the troughs. Scientists call these waves baryonic acoustic oscillations, or BAOs.

However, after those 300,000 years, the temperature of the Universe cooled sufficiently to allow atomic nuclei to soak up most of the electrons and form complete atoms.

You might say that the cosmic plasma ocean "dried up"; cosmologists call it the "epoch of recombination." Without the electrons that constantly scattered photons, light was able to pass unhindered through the Universe for the first time. This is what we detect as the CMB.

The CMB is mottled with subtle temperature variations that correspond to regions of greater and lesser density. This is the imprint of the final acoustic waves that rippled through the plasma before the epoch of recombination. The peaks of the waves mark what we describe today as the "cosmic web of matter," and it was at these peaks where galaxies, and galaxy clusters, began to form.

James Webb Space Telescope Mysteries

Illustration Only of the Big Ring

The Big Ring is centered close to 0 on the x-axis, and spans roughly from −650 to +650, equivalent to about 1.37 billion light years across.

"One possibility is that the Big Ring could be related to baryonic acoustic oscillations," says Lopez. "[These] arise from oscillations in the early Universe and today should appear, statistically at least, as spherical shells in the arrangement of galaxies.

However, detailed analysis of the Big Ring revealed it is not really compatible with the BAO explanation: the Big Ring is too large and is not spherical."

Cosmological theory suggests that the largest structures — in the form of chains of galaxies and galaxy clusters — that BAOs could form should be, at most, 1.2 billion light-years in length. Yet, the circumference of the Big Ring and the length of the Giant Arc dwarf this constraint. To put into context how immense these superstructures are, the Giant Arc is one-fifteenth the radius of the whole, visible Universe.

There are other huge superstructures in the Universe too, such as the Sloan Great Wall, which is 1.37 billion light-years across and about a billion light years away from us.

The South Pole Wall of galaxies is a more recently discovered structure; it's 1.4 billion light-years in length.

Then there is the Clowes–Campusana LQG (co-discoverer Roger Clowes is also Lopez's PhD advisor), which is a huge group of quasars spread across two billion light-years. We see these ancient quasars as they were some 9.5 billion years ago.

The Laniakea Supercluster, of which the Milky Way is a part, is tiny by comparison at just 520 million light-years across.

There are also hints of even larger structures; the "dark flow" represents the apparent motion of many galaxies in the visible Universe. This motion seems to flow in a preferred direction, as though something over the cosmic horizon were pulling the galaxies one way. However, the strength of evidence for the dark flow is controversial, with some astronomers disputing its existence in general.

Nonetheless, these superstructures are so large that not only is it difficult to understand how they formed, but it's also hard to decode how they break the Cosmological Principle, a central tenet of the Standard Model of Cosmology. This principle states that, on large scales, the distribution of matter in the Universe should be even and that no region should look substantially different from any other region. But clearly, the superstructures, and particularly the Big Ring and the Giant Arc, hugely stand out.

"Neither of these two ultra-large structures is easy to explain in our current understanding of the Universe," said Lopez. "And their ultra-large sizes, distinctive shapes and cosmological proximity must surely be telling us something important — but what exactly?"

One possibility is that the structures are hinting at exotic forms of currently known physics, or perhaps even new physics. For example the Nobel Laureate Sir Roger Penrose, who is a professor emeritus at the University of Oxford, has suggested a model called Conformal Cyclic Cosmology to describe a cyclical Universe. Per this model, evidence for gravitational waves from previous eons of the Universe could manifest as giant ring-shaped structures in the CMB. Penrose's model has not proven popular among cosmologists, but could the Big Ring and Giant Arc give it a worthy shot?

Another possibility is that the superstructures are evidence for cosmic strings, which are hypothetical one-dimensional defects in space-time believed to have formed during the Big Bang. Cosmic strings could potentially stretch across billions of light-years, yet be narrower than the width of a proton. It has been suggested that if cosmic strings exist, they could affect the clustering of matter.

"The Big Ring and the Giant Arc, both individually and together, gives us a big cosmological mystery as we work to understand the Universe and its development," concluded Lopez.

James Webb Space Telescope Mysteries

7.0 Other Space & Science Anomalies

There are more observations in space which undercut current theories of the creation of our Universe. Here some examples. The oldest star is in 7.1 and other anomalous observations are in 7.2

7.1 A Star older than the Universe

The oldest star in the Universe is HD140283 — or Methuselah as it's commonly known. This Digitized Sky Survey image shows Methuselah star, located 190.1 light-years away.

Astronomers refined the star's age to about 14.3 billion years (which is older than the Universe), plus or minus 800 million years.

In 2000, scientists looked to date what they thought was the oldest star in the Universe. They made observations via the European Space Agency's (ESA) Hipparcos satellite and estimated that HD140283 — or Methuselah as it's commonly known — was a staggering 16 billion years old.

James Webb Space Telescope Mysteries

Such a figure was rather baffling. After all, the age of the Universe — determined from observations of the cosmic microwave background — is 13.8 billion years old, so how can a star be older than the Universe?

"It was a serious discrepancy," says astronomer Howard Bond of Pennsylvania State University. So with that in mind, Bond and his colleagues set out to discover the truth and test the accuracy of the figure. Their conclusions were just as mind-blowing.

Astronomers began observing Methuselah — named in reference to a biblical patriarch who is said to have died aged 969, making him the longest-lived of all the figures in the Bible — more than 100 years ago. The curious star is located some 190 light-years away from Earth in the constellation Libra and it rapidly journeys across the sky at 800,000 mph (1.3 million kilometers per hour).

But could Methuselah really be more than two billion years older than its environment? Surely that is just not possible. Either the star was older than the Universe or the Universe was not as "young" as scientists thought it to be. Or maybe the dating was simply all wrong. What was it to be?

Investigating the Age of Methuselah

A mystery of this magnitude could not be ignored so Bond and his colleagues attempted to unearth the truth by pouring over 11 sets of observations that had been recorded between 2003 and 2011.

These observations had been made by the Fine Guidance Sensors of the Hubble Space Telescope, which noted the positions, distances and energy output of stars. In acquiring parallax, spectroscopy and photometry measurements, the scientists could determine a better sense of age.

"One of the uncertainties with the age of HD 140283 was the precise distance of the star," Bond said. "It was important to get this right because we can better determine its luminosity and, from that, its age — the brighter the intrinsic luminosity, the younger the star.

"We were looking for the parallax effect, which meant we were viewing the star six months apart to look for the shift in its position due to the orbital motion of the Earth, which tells us the distance."

Bond adds that there were also uncertainties in the theoretical modeling of the stars, such as the exact rates of nuclear reactions in the core and the importance of elements diffusing downwards in the outer layers. So they worked on the idea that leftover helium diffuses deeper into the core, leaving less hydrogen to burn via nuclear fusion. With fuel used faster, the age is lowered.

"Another factor that was important was, of all things, the amount of oxygen in the star," Bond said. HD 140283 had a higher than predicted oxygen-to-iron ratio and, since oxygen was not abundant in the Universe for a few million years, it pointed again to a lower age for the star.

As a result of all of this work, Bond and his collaborators estimated HD 140283's age to be 14.46 billion years. It was a significant reduction on the 16 billion previously claimed but it was still more than the age of the Universe itself.

James Webb Space Telescope Mysteries

<u>Investigating the Age of the Universe</u>

On the one hand, Bond says the efforts to date Methuselah is "an amazing scientific achievement which provides very strong evidence for the Big Bang picture of the Universe". By showing similarities between the age of the Universe and that of this old nearby star, he says the problem with the age of the oldest stars is far less severe than it was in the 1990s when the stellar ages were approaching 18 billion years or, in one case, 20 billion years. "With the uncertainties of the determinations, the ages are now agreeing," Bond said.

Yet, on the other hand, Matthews believes the problem has not yet been resolved. Astronomers at an international conference of top cosmologists at the Kavli Institute for Theoretical Physics in Santa Barbara, California, in July 2019 were puzzled over studies that suggested different ages for the Universe. They were looking at measurements of galaxies that are relatively nearby which suggest the Universe is younger by hundreds of millions of years compared to the age determined by the cosmic microwave background.

Far from being 13.8 billion years old, as estimated by the European Planck space telescope's detailed measurements of cosmic radiation in 2013, the Universe may be as young as 11.4 billion years. If that is, indeed, the case, then Methuselah is one again older than the Universe. The plot, indeed, thickens, but how accurate are these re-estimates proving to be?

One of those behind the studies to date the Universe is Nobel laureate Adam Riess of the Space Telescope Science Institute in Baltimore, Maryland.

James Webb Space Telescope Mysteries

The conclusions are based on the idea of an expanding Universe, as shown in 1929 by Edwin Hubble. This is fundamental to the Big Bang — the understanding that there was once a state of hot denseness that exploded out, stretching space. It indicates a starting point that should be measurable, but fresh findings are suggesting that the expansion rate is around 10% higher than the one suggested by Planck.

A higher value for the Hubble Constant indicates a shorter age for the Universe. A constant of 67.74 km per second per megaparsec would lead to an age of 13.8 billion years, whereas one of 73, or even as high as 77 as some studies have shown, would indicate a Universe age no greater than 12.7 billion years.

It's a mismatch that suggests, as stated, that HD 140283 could still be older than the Universe. It has also since been superseded by a 2019 study published in the journal Science that proposed a Hubble Constant of 82.4 — suggesting that the Universe's age is only 11.4 billion years. Astronomers are hoping the James Webb Space Telescope could shed light on this particular mystery.

Matthews believes the answers lie in greater cosmological refinement. "I suspect that the observational cosmologists have missed something that creates this paradox, rather than the stellar astrophysicists," he said, pointing to the measurements of the stars being perhaps more accurate.

"That's not because the cosmologists are in any way sloppier, but because the age determination of the Universe is subject to more and arguably trickier observational and theoretical uncertainties than that of stars."

James Webb Space Telescope Mysteries

Why Does the Universe Appear Younger than Methuselah?

But what could be making the Universe potentially appear younger than this particular star?

"There are two options, and the history of science suggests that in such cases the reality is a mix of both," Matthews said. "In this case that would be sources of observational error that haven't been fully understood, plus some gaps in the theory of the dynamics of the Universe, such as the strength of dark energy, which has been the prime driver of the cosmic expansion for many billions of years now."

He suggests the possibility that the current "age paradox" reflects time variation in dark energy, and thus a change in the rate of acceleration — a possibility theorists have found might be compatible with ideas about the fundamental nature of gravity, such as the so-called causal set theory. New research into gravitational waves could help to resolve the paradox, Matthews said.

To do this, scientists would look at the ripples in the fabric of space and time created by pairs of dead stars, rather than relying on the cosmic microwave background or the monitoring of nearby objects such as Cepheid variables and supernovae to measure the Hubble Constant — the former resulting in the speed of 67 km per second per megaparsec and the latter in 73.

Trouble is, measuring gravitational waves is no easy task, given they were only directly detected for the first time in 2015. But according to Stephen Feeney, an astrophysicist at the Flatiron Institute in New York, a breakthrough could be made over the next decade. The idea is to collect data from collisions between pairs of neutron stars using the visible light these events emit to figure out the speed they are moving relative to Earth. It also entails analyzing the resulting gravitational waves for an idea of distance — both of which can combine to give a measurement of the Hubble Constant that should be the most accurate yet.

The mystery of the age of HD 140283 is leading to something bigger and more scientifically complex, altering the understanding of how the Universe works.

"The most likely explanations for the paradox are some overlooked observational effect and/or something big missing from our understanding of the dynamics of the cosmic expansion," Matthews said. Precisely what that "something" is, is sure to keep astronomers challenged for some time.

James Webb Space Telescope Mysteries

James Webb Space Telescope Mysteries

7.2 Other Collected Observations about the Universe

The contradictions between Big Bang theory predictions and observations are not at all limited to those that have been widely dubbed a "Crisis in Cosmology".

Despite the continuing popularity of the theory, essentially every prediction of the Big Bang theory has been increasingly contradicted by better and better data, as shown by many teams of researchers.

The observations are, on the other hand, consistent with a non-expanding Universe with no Big Bang. The real crisis in cosmology is that the Big Bang never happened.

Key contradicted predictions are detailed below:

1) Light elements: Lithium and Helium

Prediction: Any superhot explosion throughout the Universe, like the Big Bang, would have generated a certain small amount of the light element lithium and a large amount of helium.

Observation: Yet as astronomers have observed older and older stars, the amount of lithium observed has gotten less and less, and, in the oldest stars is less than one tenth of the predicted level. The oldest stars near to us have less than half the amount of helium predicted. However, well-understood fusion processes in stars and reactions initiated by cosmic rays have accurately predicted the correct amounts of these and other light elements.

James Webb Space Telescope Mysteries

2) Antimatter-matter annihilation

Prediction: Since the intense radiation of the Big Bang would produce matter and antimatter in equal amounts, mutual annulation of particle-antiparticle pairs would reduce the surviving matter density to around 10^{-17} protons/cm3.

Observation: the matter density in the Universe is observed to be at least 10^{-7} ions /cm3 more than 10 billion times higher than the Big Bang prediction.

Big Bang fix to prediction: To try to fix this well-known vast gap, Big Bang theorists have proposed some unknown asymmetry between matter and antimatter which would lead to more production of matter. This has never been observed in laboratory experiments. A consequence of this predicted imbalance is the decays of the proton, initially predicted to decay with a lifetime of 10^{30} years. Large scale experiments have contradicted this prediction was well, with no evidence of decay at all.

3) Surface-Brightness

Prediction: In any expanding Universe, an optical illusion makes objects at high redshift appear larger and dimmer, so their surface brightness—the ratio of apparent brightness to apparent area—declines sharply with redshift.

Observation: Based on observations of thousands of galaxies, surface brightness is completely constant with distance, as expected in a Universe that is NOT expanding.

James Webb Space Telescope Mysteries

Big Bang fix to Prediction: After observations showed that the surface brightness dimming did not occur, Big Bang theorists hypothesized that galaxies were much smaller in the distant past and have grown greatly. But observations have contradicted this fix as well, showing that there have not been enough galaxy mergers for the growth rates needed. In addition, the ultra-small galaxies hypothesized would have to have more mass in stars than total mass, an obvious impossibility.

4) Too Large Structures

Prediction: In the Big Bang theory, the Universe is supposed to start off completely smooth and homogenous. Structure starts small and grows over time

Observation: As telescopes have peered farther into space, huger and huger structures of galaxies have been discovered, which are too large to have been formed in the time since the Big Bang.

5) Cosmic Microwave Background Radiation (CMB) and its Anisotropies

Prediction(Initial): The CMB is a smooth relic of the initial radiation of the Big Bang.

Observation: The CMB is smooth on such large scales that , in a Big Bang there would be too little time for regions that we now see in different parts of the sky to reach equilibrium with each other, or even to receive energy from each other at the speed of light.

Big Bang fix to prediction: An unknown force, dubbed "inflation" generated an exponential phase of the Big Bang that blew up the Universe so rapidly that all asymmetries were smoothed away.

Additional observations: The actual very small anisotropies in the CMB were much smaller than those predicted by Big Bang theorists and additional fixes had to be added to the theory each time the observations became more precise, so that at present seven free variables—the density of dark matter, of ordinary matter, of dark energy and four additional fitting parameters—are needed to fit the observations. They still badly fail with some of the largest-scale anisotropies

The latest crisis: Based on the data from the Planck satellite, the best fit to the CMB predicts a Hubble constant (the ratio of redshift to distance) in conflict with observations based on Supernovae.

The best fits imply a curved Universe, in conflict with the predictions of inflation for a flat Universe. And they predict a density of dark matter far greater than any measurements derived from the motion of galaxies.

In contrast to the multiple contradictions of the Big Bang theory of the CMB with its "ultra precise" but wrong predictions, non-Big Bang processes provide a better explanation. The energy that was released in producing the observed helium in the Universe equal the energy in the CMB. Any radiation become isotropized if it travels in a medium that scatters it. There is abundant observational evidence that microwave-frequency radiation is scattered in the intergalactic medium.

6) Dark Matter

James Webb Space Telescope Mysteries

Prediction: The Big Bang theory requires the existence of dark matter—mysterious particles that have never been observed in the laboratory, despite huge experiments to find them.

Observation: Multiple lines of evidence, especially observations of the motions of galaxies, show that this dark matter does not exist. Extremely sensitive experiments on Earth have failed to detect dark matter particles. In addition, dark matter, if it existed would create a viscosity effect on galaxies that would prevent the existence of the many long-lived groups of galaxies that are observed.

The response of most cosmologists to this growing body of evidence has, unfortunately, not been to decide the Big Bang theory has been falsified, but to add new "parameters" and hypotheses, like dark energy.

The theory is now far more complex and speculative than the Ptolemaic epicycles that were destroyed by the Scientific Revolution. Each contradiction with observation is taken as a mere "anomaly" that does not undermine the theory as a whole. Strong peer pressure is applied against many of those who question the theory.

"It's as if researchers are saying 'I can see the Emperor's elbow through his New Clothes,' 'I can see the Emperor's knee though his New Clothes' and so on," says Lerner. "It is time to say: 'The Emperor is not wearing any clothes.' This theory has no correct predictions."

To replace the Big Bang, other researchers have elaborated, in peer-reviewed publications, alternative explanations of the generation of light elements and of the energy in the CBR by ordinary stars, and of the development of large-scale structures through the interaction of gravity and electromagnetic processes. "No one would claim that all the problems in cosmology have been resolved," agrees Lerner, "but the evidence is consistent with an evolving, but non-expanding Universe, which had no beginning in time and no Big Bang."

James Webb Space Telescope Mysteries

8.0 Possible Explanations

There are numerous explanations about the anomalies discovered by the James Webb Space Telescope (JWST) and how they affect our theories of the Universe.

Other observations also need explanations. Here are some of those explanations.

8.1   The Universe Already Existed

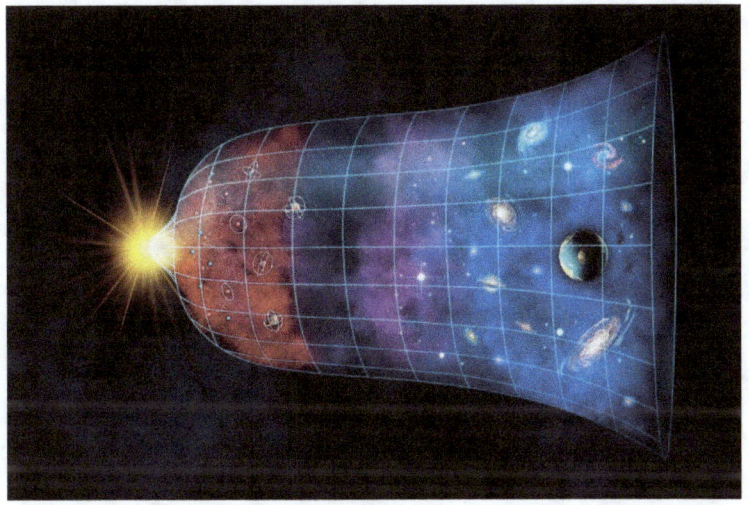

Space-Time is just like what Einstein mentioned some decades back that the distinction between pasta present and future is only an illusion.

No matter how persistent they might look is time Universal. In other words is there an audible Tick Tock throughout the Galaxy a Master Clock so to speak.

The answer my friends is no time is not absolute so let's recap where we stand now. Penrose says that the Universe restarts endlessly from the big explosion to the next round. But it means it is even harder to track the beginning or ending of The Big Bang in this cyclic model and so in this intricate Cosmic Landscape we catch a glimpse of the Intriguing possibility where our Universe is constantly growing and creating new Universes within itself.

Most of space would be this stuff which is continuing to inflate and then it would produce Quantum fluctuations that continue the inflation further while maybe producing other patches where inflation ends and this would continue literally eternally and we call this **Eternal Inflation** when combining quantum physics with gravity and introducing inflation. The outcome is eternal.

Inflation in this scenario most of space is always inflating with pockets where inflation has ceased and due to Quantum fluctuations.

Some of these areas resemble our Universe while others don't. An infinite number might appear like ours but an equally infinite number will not. This leads to the idea of a Multiverse where every conceivable event or scenario will occur infinitely given the Universe's physical ingredients.

With such infinite possibilities the probability of a patch resembling our Universe is astronomically low

Recent events have cast doubt on some aspects of this idea. In March 2014 the bicep 2 group from Antarctica claimed to have detected Cosmic gravitational waves some touted as the conclusive proof of inflation. However it was later revealed that the observed waves were not Cosmic in origin but patterns created by light passing through the dust of our galaxy.

James Webb Space Telescope Mysteries

Instead of proving inflation this discovery raised more questions. Recent findings have set stringent limits on gravitational waves excluding many simple inflationary theories. This led to a renewed focus on the foundational assumptions of The Big Bang Theory.

Some theorists are now revisiting the idea of the big bounce an alternative to the Big Bang. In this model the Universe is smoothing and flattening. It doesn't result from a high energy random Quantum origin.

Instead it emerges during a time resembling our present where classical physics dominates. This model proposes that the Universe undergoes periods of slow contraction instead of rapid expansion eliminating the Multiverse concept caused by random Quantum fluctuations during inflation.

The big bounce Theory also has its predictions. It suggests a Universe devoid of cosmic gravitational waves because of the absence of energy concentration. Multiple experiments are ongoing to determine the presence or absence of these waves.

Another significant component of this theory is the idea of a cyclic Universe where the Universe undergoes periodic bounces. This proposition is currently being tested through experiments and observations. The findings from these studies will not only determine the origins and future of the Universe but will also establish the relationship between gravity and quantum physics.

These discoveries will shape our understanding of the Universe's nature. If the Multiverse theory is correct our observable Universe might just be a random occurrence detaching us from understanding its fundamental creation.

But if the bounce theory is right our observations would represent the Universe as a whole giving insights into its fundamental laws.

These upcoming experimental results
will have profound implications on our comprehension of the cosmos.

James Webb Space Telescope Mysteries

8.2 Galaxies Formed Very Fast

Using the James Webb Space Telescope, scientists have discovered that early Universe galaxies must have grown up way faster than expected. Plus, the same team also found that, 10 billion years ago, the cosmos wasn't quite as disordered and chaotic as previously believed.

The international team, led by researchers from Durham University in the U.K., reached these conclusions by discovering evidence of structures called "star bars" forming in galaxies that existed just a few billion years after the Big Bang.

Star bars are elongated regions of increased star density found at the hearts of spiral galaxies like the Milky Way and other disk galaxies. As they form, star bars push gas toward the hearts of their respective galaxies, thereby regulating star birth. The presence of these central bar structures thus indicate that a galaxy has entered a more settled and "mature phase."

James Webb Space Telescope Mysteries

"Galaxies in the early Universe are maturing much faster than we thought," Zoe Le Conte, team leader and a researcher at Durham University, said in a statement. "This is a real surprise because you would expect the Universe at that stage to be very turbulent, with lots of collisions between galaxies and a lot of gas that hasn't yet transformed into stars.

"However, thanks to the JWST, we are seeing a lot of these bars much earlier in the life of the Universe, which means that galaxies were at a more settled stage in their evolution than previously thought. This means we will have to adjust our views on early galaxy evolution."

Bar-hopping for the James Webb Space Telescope

This isn't the first time scientists have gone bar-hopping in the early history of the 13.8 billion-year history of the Universe.

The Hubble space telescope witnessed these features as well, but that orbiting eye on the Universe could only go as far back as 8 billion to 9 billion years. The increased sensitivity and wavelength range of the JWST, however, has stretched such observations back at least another 1 billion years.

This has revealed bar formation in galaxies that are seen as they were between 8 billion and 11.5 billion years ago. In fact, of 368 disk galaxies the team considered for the study, 20% already had bars.

That is double the number observed by Hubble.

8.3　We Really Live in a Multiverse

Can we really cling to a 100-year-old set of Einstein's formulas when there is real evidence that these foundations of our physics are as incomplete as Newton's theory of gravity was.

For example let's consider alternative theories for a moment. There is enough evidence in the quantum world that the Multiverse is more likely than a single existing Universe that has a beginning and an end.

James Webb Space Telescope Mysteries

It's much more likely that our Universe is just one of many and that there are other dimensions in which completely different basic physical rules apply;

Sir Roger Penrose found evidence that the Universe is cyclical it comes into being exists passes away and then comes into being.

Again other ideas suggest that our Universe is just a kind of bubble in a much larger Universe. If this is true we would have to reach the edge of our bubble at some point and then a new space would begin.

This may be the case now the pocket Universe for which there is also good evidence is designed in such a way that there are different plots within a space which are like Pockets whose exits lead to other dimensions.

James Webb Space Telescope Mysteries

9.0 Summary

After researching what science and astronomy has learned about the Universe in recent years and various explanations scientists have for the various anomalies, I have come to my own conclusions.

I think our "Big Bang" occurred into an existing Universe and other Universes may have already existed which are separate but which ours may bump up against.

These other Universes also have different physical laws which make them appear a lot different. We actually live in a Multiverse.

I also have some possible evidence from other research of mine into Irish mythology. The Tuatha Da Dannon were the forerunner race of beings who later became leprechauns and Druids.

They said they came from the "Otherworld" which was another place they could access through dimensional portals. Humans who visited the Otherworld and returned found that weeks spent there might have become years here on Earth while they were gone.

This statement about different rates of time is what you would expect-that different Universes have different physical properties such as time running at a different time rate. I also have other evidence that parallel worlds exist. Please see my book "Stories of Parallel Dimensions".

Also, how could the size of the known Universe be a sphere with a diameter of 90 billion light years when given the time since the Big Bang occurred there would not be time for the Universe to expand to this size given the speed of light.

Of course scientists have an explanation for this inflation saying that the properties of the early Universe were a lot different and the Universe expanded to its current size at a rate much greater than the speed of light. So there would have to be a lot of changes in the physical properties of our Universe to make this expansion rate true.

A much simpler explanation is that the Universe existed long before the last "Big Bang" and has had plenty of time to expand to its current size of a 90 billion light year diameter.

The time and space continuum we live in then is at least forty five billion years and possible much older. We just can't see the true edge of it.

James Webb Space Telescope Mysteries

10.0 Bibliography

https://www.youtube.com/watch?v=DAnwll1jYWE. (2024). Retrieved from James Webb Telescope Proves Big Bang Wasn't The Beginning of Everything, Something Exists Beyond It.

https://www.youtube.com/watch?v=L2nhuskEcyk. (2024). Retrieved from James Webb Telescope Saw 88 Strange Galaxies.

James Webb Telescope Saw First Object Beyond the Dark Ages. (2024). Retrieved from https://www.youtube.com/watch?v=eZPng4BynpY.

JWST Instruments. (2024). Retrieved from https://webbtelescope.org/news/webb-science-writers-guide/webbs-scientific-instruments.

New Hypothesis Explains How the Big Bang May Never Have Occurred! (2024). Retrieved from https://www.youtube.com/watch?v=2yX0SKx9iol.

New Image from The JWST Shows That Something Is Seriously Wrong with Our Universe. (2024). Retrieved from https://www.youtube.com/watch?v=U7-3AmR6w_0.

Science News--Proto Clusters. (2024). Retrieved from https://www.sci.news/astronomy/webb-proto-globular-clusters-early-Universe-13043.html#:~:text=Astronomers%20using%20the%20NASA%2FESA,was%20460%20million%20years%20old.

The Scientific Evidence Against The Big Bang. (2024). Retrieved from https://www.lppfusion.com/science/cosmic-connection/plasma-cosmology/the-growing-case-against-the-big-bang/.

Video About the Multiverse. (2024). Retrieved from https://www.youtube.com/watch?v=Z3pa9DfmJGQ.

www.ingramcontent.com/pod-product-compliance
Lightning Source LLC
Chambersburg PA
CBHW071950210526
45479CB00003B/877